W9-AAG-791

Michael Faraday's Candle

BURN

by Darcy Pattison

Illustrated by Peter Willis

BURN: Michael Faraday's Candle

Printed in the United States and United Kingdom..
For permissions contact:
Mims House
1309 Broadway
Little Rock AR 72202
USA
MimsHouse.com

Publisher's Cataloging-in-Publication data

Pattison, Darcy.
    Burn : Michael Faraday's Candle / by Darcy Pattison ;
illustrated by Peter Willis.
    pages cm
    ISBN 978-1-62944-044-6 (Hardcover)
    ISBN 978-1-62944-045-3 (pbk.)
    ISBN 978-1-62944-046-0 (ebook)
    Summary : Adaptation of Michael Faraday's lecture
explaining why a candle burns.

1. Faraday, Michael, 1791-1867 --Juvenile literature. 2. Chemistry
--Juvenile literature. 3. Candles --Juvenile literature. I. Willis, Peter
N. II. Title.

QD39 .P38 2015
[540] --dc23   2015912911

Adapted from Michael Faraday's "Chemical History of a Candle"

BURN

by Darcy Pattison

Illustrated by Peter Willis

It was 3 p.m. on December 28, 1848 in London, England.

Horses whinnied. Excited boys and girls stepped out of carriages into the crowded one-way Albemarle Street and pushed into the Royal Institution. They were excited; this was a day that they would not talk about science. Instead, they would see science.

THE ROYAL INSTITUTION OF GREAT BRITAIN

Mr. Michael Faraday, Director of the Laboratory, would be giving the children's Christmas lecture. It cost one guinea to attend.

About 4000 people crowded onto hard wooden benches. The three-story auditorium was stacked so everyone in the audience could see the experiment desk. In previous years, some experiments turned fiery. People sat back, not too close.

MICHAEL FARADAY'S
CHRISTMAS LECTURE

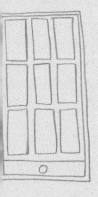

Mr. Faraday stepped up to the lecture table and began:

"I bring before you the Chemical History of a Candle."

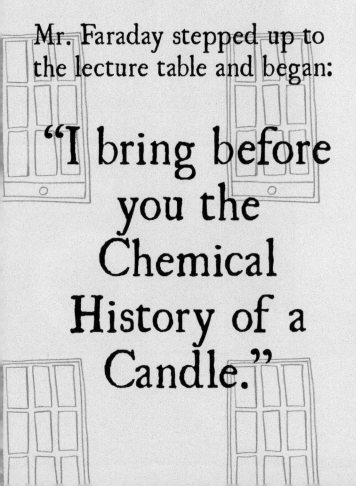

"I must first tell you of what candles are made.

The candle I have in my hand is a stearin candle, made of ox fat.

Then here is a sperm candle, which comes from the purified oil of the sperm whale.

Here, also are yellow and refined bees-wax, from which candles are made.

Here, too, is that curious substance called paraffine, and some paraffine candles."

66

"Now as to the light of the candle, we will light one or two, and set them at work.''

"The candle is a solid. And how is it that this solid can get up to the place where the flame is?

This is a wonderful thing about a candle."

"You see that a beautiful cup is formed. As the air comes to the candle, it moves upward because of the heat of the flame, and so it cools the sides of the wax. The cup is formed by the air rising which keeps the outside of the candle cool."

What is the cause

"I hope you will always remember that whenever a result happens, especially if it be new, you should say, 'What is the cause? Why does it occur?'

And you will, in the course of time, find out the reason."

"But how does the flame get hold of the fuel? There is a beautiful point about that—capillary action."

"'Capillary action!' you say—well, never mind the name.

It is by what is called capillary action that the fuel is carried to the part where the combustion or burning goes on."

Capillary action happens when two things won't dissolve in each other, but instead hold together.

MADE IN ENGLAND

"When you wash your hands, you take a towel to wipe off the water. Capillary action makes the towel become wet with water. In the same way, a candle's wick is made wet with the wax.

The melted wax climbs the cotton wick to get to the top; other wax particles follow because the wax particles are attracted to each other. As the wax particles reach the flame, they are gradually burned."

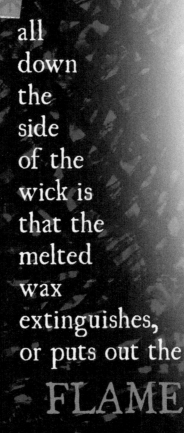

"The only reason why the candle does not  **BURN** all down the side of the wick is that the melted wax extinguishes, or puts out the FLAME."

"Now as to the shape of the flame, it is a bright oblong, brighter at the top than toward the bottom, with the wick in the middle, and certain darker parts toward the bottom."

MICHAEL FA

HRISTMAS

There is a current formed which draws the flame out. You may see this by taking a lighted candle and putting it in the sun so as to get its shadow thrown on a piece of paper. You see streaming upward the current of hot air, which draws out the flame, supplies it with air, and cools the sides of the cup of melted fuel."

"It is too bad," said Mr. Faraday, "that we have not gotten farther, but we must not keep you beyond your time."

English 75

in London.

T

THE END

MICHAEL FARADAY (1791-1867)

British scientist and science educator Michael Faraday was originally apprenticed to a bookbinder, George Riebau. Soon, however, he was less interested in binding books and more interested in reading the books about science. By 1809, he started keeping a journal of his own experiments, attending lectures by other scientists, and writing books about science. By 1813, Faraday was employed as a laboratory assistant by the Royal Institution, where he worked for most of his life.

Faraday is best known as an experimenter. He always asked, "What is the cause? Why does it occur?" And then he created experiments to answer the questions. He's most famous for his work in electro-magnetic rotations, which is the basis of the electric motor. In chemistry, he discovered two elements, chlorine and carbon. Some of his work took gases such as benzene or chlorine and converted them to liquids. More practical for his time period, he experimented with steel alloys and optical quality glass. In the laboratory, he invented an early form of the Bunsen burner, a convenient source of heat for experiments.

While he was well-known as a scientist, Faraday was also a man of faith. A devout Christian, he served as a deacon and elder in his church. In his personal life, Faraday married Sarah Barnard. They had no children. See more of his biography at http://www.rigb.org/our-history/michael-faraday/about

THE ROYAL INSTITUTION'S CHRISTMAS LECTURE SERIES
1825-1938, 1943-PRESENT http://www.rigb.org/christmas-lectures

Since 1825, the Royal Institution has presented a Christmas lecture for the "juvenile" scientists, or for children. Since then, it has run continuously except during World War II. Over the years, almost every area of science has been discussed: biology, chemistry, astronomy, physics, robotics and much more. Still popular today, the Royal Institution Christmas lectures are the longest running series of science education lectures in the world.

Michael Faraday presented 19 lectures between 1827 and 1860. He gave his most popular lecture first in 1848 and repeated it in 1860, "The Chemical History of a Candle." Since its publication in 1861, it's never been out of print. This book recreates the first section of the lecture by carefully adapting the 6500-word original text to about 650 words.

OBSERVE THE CHANGES: SOLID, LIQUID, GAS

When wax from a candle is burned, the wax changes from a solid to a liquid. The wax cup just below the flame holds the melted, liquid wax. This change is reversible. If you blow out the candle, the wax in the cup will cool off and become solid again. However, when the candle is lit, and the wax is burned, the chemical changes are not reversible. You can never put the wax back into the candle.

NGSS STANDARDS

1st Grade: PS4-2 to 4-4. Waves: Light; 2nd Grade: PS1-1 to PS1-4 Matter and its interactions; 4th grade PS3-2, PS3-4 Energy; 5th Grade: PS1-1 to PS1-4 Matter and its interactions

READ MORE

The National Candle Association provides a safety poster, science projects, and more at candles.org.

CANDLE WAX

Candle wax doesn't occur naturally. Chemical and mechanical processes are applied to the matter to change it to candle wax. In his lecture, Faraday described how waxes were purified.

OX FAT: The fat, or tallow, is first made into a soap, and then the soap is treated to make stearic acid and glycerin. The oil is then pressed out of it until only stearin wax remains.

SPERMACETI OIL: Oil is taken from a head cavity of a sperm whale. The oil is put under pressure and then treated with chemicals to produce spermaceti, a white crystal of wax.

BEESWAX: Beeswax comes from the honeycomb. After the honey is drained, the wax is melted and refined.

PARAFFIN: In his lecture, Faraday showed examples of candles made from paraffin (or paraffine, as it was spelled in Faraday's time). Yet, paraffin had only been discovered a year earlier in 1847. At the time, it was made from peat from Irish bogs. Paraffin candles couldn't be bought in a shop until 1858. To make paraffin, peat was carefully heated until paraffin flakes were created. It was then treated with chemicals, heat and pressure to purify it.

CPSIA information can be obtained
at www.ICGtesting.com
Printed in the USA
LVOW05*1406100716

495758LV00042B/354/P